M. de Luc

Account of a New Hygrometer

By M. J. A. De Luc, Citizen of Geneva, F. R. S. and Correspond. Member of

the Academies of Paris and of Montpellier

M. de Luc

Account of a New Hygrometer
By M. J. A. De Luc, Citizen of Geneva, F. R. S. and Correspond. Member of the Academies of Paris and of Montpellier

ISBN/EAN: 9783337382124

Printed in Europe, USA, Canada, Australia, Japan

Cover: Foto ©berggeist007 / pixelio.de

More available books at **www.hansebooks.com**

XXXVIII. *Account of a new Hygrometer.* *By* M. J. A. De Luc, *Citizen of* Geneva, *F. R. S. and Correspond. Member of the Academies of* Paris *and of* Montpellier.

Read June 10, 1773.

IN laying before the Royal Society an account of my attempts to find out a method for measuring the moisture of the air, I think myself obliged to relate the gradual steps of my mind, the obstacles I met with, the means by which I endeavoured to overcome them, the degree at which I flatter myself to have arrived, the hopes that may be entertained of farther advances, and the uses which may be derived from my first experiments.

Attempts to invent an HYGROMETER.

1. In order to proceed regularly in this investigation, I began by examining the essential requisites in a machine intended to measure humidity, which I found to be the three following:

1st, The settling of a fixed point, from which every measure of the same kind should be taken, such, for instance, as that of boiling water in a thermometer, when the barometer is at a certain height.

2d, Degrees equally determined, or comparable, in different hygrometers, such as are in the thermometer,

meter, the scales of Fahrenheit, Delisle, Reaumur, &c.

3d, Constancy in the variations produced by the same differences of humidity.

2. I perceived, moreover, that it were to be wished, that the hygrometer should give a true indication of the relation between the real quantities of the humidity, or at least between their differences: but this last point I rather considered as a desirable degree of perfection, than as an indispensable requisite; the essential point being, that observers might understand each other, when mentioning degrees of humidity; and this seemed to be sufficiently provided for by the abovementioned conditions.

3. Having thus planned to myself the work I had to go through, I first attended entirely to the first point, and laid aside all the others. This I again subdivided. I had soon perceived that I must begin by thinking much less of the hygrometer than of the different phænomena of humidity. For this purpose it was necessary to find out a fixed state, either of bodies in general, or of some body in particular; and this fixed state might either be extreme humidity, or dryness, or any intermediate point.

4. Knowing that the extremes in nature are commonly very difficult, and sometimes impossible to hit, I conceived at first greater hopes of intermediate degrees. But in vain did my imagination fatigue itself in a road, which I was forced to abandon.

5. I then came to the extremes, and that of absolute dryness was the first I was induced to try. But having found no other way to procure it but by fire, and fire not producing it in all bodies which appeared

to

to me fufceptible of humidity, but by altering their nature, I reluctantly perceived that I fhould be obliged to look for my firft point, where I had the leaft hopes of difcovering it.

6. I remained a long time without difcovering any thing in this new road ; and very often turned back, but was always obliged to return to extreme humidity, as to the only part of my object, of which I could poffibly get any hold.

7. The words, which are neceffary for communicating our ideas to others, are often obftacles to the raifing of new ideas in ourfelves. They are by far too few to exprefs diftinctly every fhade of intellectual objects. Humidity was a word which I conftantly repeated to myfelf, and it conftantly led me to a clafs of phænomena, in which I could find nothing fettled.

8. Water at length prefented itfelf to my mind ; and in this fluid, which to all appearance ought firft to have ftruck me, I beheld with furprize, what I had been labouring, through many a round, to difcover, under the denomination of extreme humidity. I was not at that time confidering humidity in any particular phænomenon ; I only obferved that it was conftantly produced by aqueous particles diffeminated through bodies ; and I found in water the maximum of the approach, and confequently of the action, of thefe particles.

9. In order now to avoid the ambiguities from whence, in my opinion, the difficulties in thefe matters arife, let me be allowed for the future to employ no words but fuch whofe meaning is well determined. Humidity will accordingly be no more
that

than an effect, or modification of bodies from a fubftance more or lefs abundant, but conftantly confift-
ing of aqueous particles under different forms. This
fubftance, confidered in its utmoft extent and under
all the appearances which it affumes in nature, I
fhall exprefs by the Latin word *humor*. Thus ice,
water in its different degrees of heat, hail, fnow,
icicles, rain, dew, clouds, fog, mift, invifible vapours,
are no more than modifications of this fame fubftance, different fpecies of a determined genus; fince
aquofity, which is common to all, is its generic
character.

10. The more humor there is in any body, the
more humid that body is; and confequently if it be
plunged in water, and foaked fo as not to be able
to receive any more, it is got to extreme humidity,
the water which fills up all its pores being humor in
the higheft degree of intenfity.

11. Not, however, but that difcrete humor, or
vapour of every kind, may in fome refpects produce
as great effects as concrete humor or water: but
there is always fome difference in fome other refpect,
and chiefly in regard of time. Bodies encompaffed
with air are continually difcharging, by evaporation,
part of the humor they imbibe from it. If the circumftances are fuch, that the humectation exceeds
the evaporation, the body at length wets through *,
more or lefs quickly as the quantity of humor which
it receives in a given time is greater or lefs, and likewife in proportion as this quantity exceeds that

* By *wetting* here, I underftand arriving at the greateft degree of humidity.

which

which evaporates. It is fuddenly wetted, when the humor is fo condenfed as to become water, becaufe the evaporation which takes place at the furface of the water, does not weaken its action on the bodies dipped into it; it is only wetted little by little, or, what happens ofteneft, in part, when the humor is difcrete or reduced into vapour; becaufe, while it is depofited on particular fpots, it evaporates from the interftitial parts, and that more or lefs according to the ftate of the air, and that of the moiftened bodies.

12. This difference, however, in point of time, between the action of concrete and difcrete humor, only takes place on the furface of bodies, or at a fmall depth ; it diminifhes, and may even become oppo-fite, as the depth of bodies increafes, becaufe the difcrete humor is then more eafily introduced into their pores than water, which more than makes up for their different intenfity.

13. This confideration folves a difficulty, which at firft puzzled me. I had been told by bird-catchers, that the threads of thofe nets which they caft on the water-fide, were lefs ftretched from the action of water than of dew. Hence it might feem, that what I took for the extreme of humor had lefs effect than what is only a degree of it. But two particular caufes accounted for this difference.

1ft, The air contained within the fibres of the thread oppofes the introduction of the water, which, prefenting itfelf in a body, fhuts up the paffages by which the air fhould efcape to give it room; but it yields to the drops of dew, which permit its efcape while they penetrate through the threads.

2d, A ohe r

2d, Another particular caufe, lefs obvious though not lefs probable than the preceding, is the difference in the mutual attraction of parts, in the concrete and difcrete humor, and confequently in their refpective facility to feparate, and get one by one through the narrow pores. When this entrance is attempted by the humor, under the form of water, the mutual attraction of its parts, being greater than in dew, occafions a greater refiftance to their introduction, than when they are already divided by fome other caufe, viz. when the humor is reduced to fmall drops, or vapor.

14. This phænomenon, therefore, does not contradict my principle; it is only a particular fact; and it remains true, that bodies furrounded with water are expofed to the extreme of humor. To remove this caufe of exception from my hygrometer, it was fufficient to provide outlets for the air, and not to increafe too much the thicknefs of the body, upon which the humor was to act.

15. Another difficulty, which prefented itfelf, was that water might probably act with more or lefs energy in proportion to its heat. But this did not ftop me long. As my prefent object was a fixt point for the hygrometer, and not the greateft power of water, confidered as a caufe of humidity, it was enough to employ it conftantly at the fame degree of heat; and, to fix this with greater precifion, I determined to ufe water at the inftant that it ceafes to be ice. The bafis therefore of my hygrometrical fcale was to be the foaking power of melting ice.

16. This principle, being thus unfolded, appeared fo fimple, that I was at firft furprifed how it could have

have been fo long overlooked. But I afterwards accounted for this, from the difficulties which I met
with in the difcovery. The notion of an hygrometer being both complex and unfettled, all the obftacles prefented themfelves at once, and this multitude of ideas exceeded the power of attention. The
very firft fteps were apt to miflead. On the one
hand, I looked for an hygrometer with a head full
of the matters already ufed for hygrofcopes, which
always are more or lefs altered by water; and, on
the other, the name of humidity was applied to that
caufe, the effeds of which I wifhed to meafure; and
both points of view turned afide the mind from the
idea of water, as being proper to afford the required
fixed point in an hygrometer.

17. The firft difficulty had not efcaped me; but,
confidered in itfelf, it did not appear unfurmountable.
I was in hopes that a fubftance might be found capable of being affeded by the foaking power of
water, without being altered by it. As the nature
of this fubftance was to determine not only the form
of the hygrometer, but alfo the fpecies of the degrees, which were to indicate the different quantities
of humor, I concluded that my fecond objed ought
to be the difcovery of this fubftance.

18. In this refearch, I again divided the objeds,
by confidering feparately the three kingdoms, viz. the
mineral, the vegetable, and the animal. The two
firft offered no fubftance fit for my purpofe, viz.
none that would obey the impreffions of humor,
without being altered either by it or by other caufes.
But in the animal kingdom bones drew my attention; and ivory, in particular, feemed to poffefs the
                                        required

required qualities. I had obferved that the key of an ivory cock was tighter or flacker, as there was more or lefs humor in the air. Ivory pallets, ufed for water-colours, fhewed no alteration, at leaft none that was lafting. I knew alfo the elafticity of this fubftance, which feemed to fecure its coming back to the fame ftate, on its return to the fame degree of moiftnefs.

19. There ftill remained on this fecond head another object of inquiry, which was almoft neceffarily connected with the third, viz. the fpecies of the degrees to be given to the hygrometer. The beft form to be given to the ivory, in order to receive with eafe the impreffions of the humor, and to have its effects meafured upon it, was to be determined. I firft thought of ivory rods, the lengthening of which fhould be meafured by a machine fimilar to the pyrometer. I likewife had fome notion of a large nonius, formed of an ivory and a metallic rod. Either of thefe machines would admit of a fixed graduation, as both the dimenfions of their parts and proportions to one another could be determined. But then I apprehended that ivory might perhaps, like wood, have its longitudinal fibres but little liable to be extended by the humor, and that the imperfections of thefe two kinds of micrometers would occafion a confiderable irregularity in the hygrometrical degrees. I alfo feared that if ivory rods were made thick enough to prevent their bending, fuch a thicknefs might become an obftacle to their intire penetrability by the humor (14). I therefore concluded that the ivory fhould have fuch a form, that, though very thin, it might not warp; and that the meafurable varia-

tions were to be the removal or approach of its fibres to each other.

20. Being thus guided by thefe neceffary conditions, I thought of different thin ivory cups, the capacities of which fhould be meafured by quickfilver; and at laft imagined a hollow cylinder, in which the variation of its capacities, when more or lefs moift, might be meafured by the quickfilver it fhould be filled with; and which putting into a glafs tube joined to the ivory one, would of courfe rife more or lefs, as that veffel was more or lefs deprived of humor.

21. Nothing now remained but to find out a way of eftimating the changes of capacity of the ivory tube, by means of the variations in the height of the mercury in the glafs tube. I thought, at firft, that by ufing very nice fcales, in order to compare the weights of the mercury contained in the cylindrical veffel, with that of a column of the fame liquid in the tube, I might obtain the proportions of thefe weights with a fufficient exactnefs, to be able to meafure the variations of the mercurial column, by degrees reprefenting aliquote parts of the whole mafs.

22. This in itfelf was undoubtedly an exact method; but then it required in the execution fuch a nicety in the fcales, that I durft not employ it in the conftruction of an inftrument of fo extenfive an ufe. Such fcales are always fcarce from their high price. I remembered to have myfelf found that inconvenience in the conftruction of a Delifle's thermometer, and concluded I muft hit upon fome method to avoid it.

23. The idea of a thermometer, which ftruck my mind, was a lucky one. I was led to it by a
kind

kind of connexion between the scale of that inftrument and that of my hygrometer. I foon perceived,
that, by applying to my hygrometer a thermometrical tube, already graduated by means of two fixed
points of heat, it would only be neceffary to know
the proportion of the weights of mercury in this thermometer, and the hygrometer, to which its tube was
to be applied, to have in this laft inftrument degrees
as well determined as in the firft. Scales of a common degree of exaftnefs were fufficient to eftablifh
between the refpeftive degrees of both inftruments
a proportion equal to that of their mercurial weights
(42 and 43).

24. Befides the eafe in the execution, this contri-
vance afforded me a very fimple method to correft
the effefts of heat upon the mercury contained in
the hygrometer. It is indeed obvious, that, abftraft-
edly of the effefts of the humor, the new inftrument
muft in itfelf be a regular thermometer; and that
confequently the variations of an adjoint thermometer were immediately to point out this correftion.

25. Every principle being thus fettled, nothing
remained but to contrive its conftruftion. I began
by making fome experiments, on the nature and
quantity of the aftion of water upon ivory. I made
for that purpofe a fmall cylindrical ivory veffel, of
an inch in diameter, and eight lines in length, and
reduced its thicknefs to lefs than ¼ of a line. I likewife prepared a wooden cylinder, equal in its diameter to the internal one of the veffel. I then put
this veffel into water, in fuch a manner that it only
wetted it outwardly to the rim. In a very fhort
time the wooden cylinder, which at firft filled the

cup

cup exactly, no longer filled it.   After a few hours,
I perceived that the internal furface grew wet, and by
means of a magnifying glafs, found it covered with a
very fine dew.   This dew did not encreafe by the
veffel remaining any longer in the water ; the evaporation being doubtlefs equal to the tranfudation ;
and the capacity of the veffel, which encreafed till the
appearance of the dew, feemed afterwards at a ftand.

26. This tranfudation puzzled me a little ; it
fhewed me that the water would get into my hygrometer, which at firft appeared an inconvenience.   I
foon, however, found an advantage in it.   The water, after having foaked through the ivory, would
immediately pufh back the mercury, which having
by degrees funk in the tube, during the penetration
of the water through the pores of the ivory, muft
thus rife again.   Hence I might expect a maximum
for the fall of the mercury, very eafy to be determined.   As for the water introduced into the ivory
veffel, I was in hopes that it would go back as foon
as the outfide of the cup fhould be dry.

27. Having thus afcertained that ivory was very
eafily affected by the impreffions of the humor, it ftill
was neceffary to know, whether the variations of the
one would always equally anfwer to thofe of the
other.   Having accordingly taken my fmall cup out
of the water, and expofed it to the air, I foon found
that its capacity diminifhed, but that even after feveral days it did not return to its former ftate.   This
again puzzled me ; but I fufpected that the external
preffure of the tool upon the ivory might fomewhat
have compreffed it, and that the water having re-
ftored the ivory fibres to their original pitch, the
abfolute

absolute capacity of the cup remained larger than it was before.

28. To satisfy myself about this, I got another wooden cylinder, which filled the capacity of the veffel in its prefent ftate. This I again put into the water, and left it there a fufficient time; I then expofed it to the air to be dried; and after that found that the wooden cylinder filled it as before. Hence I concluded, that in the conftruction of my hygrometer the ivory cup fhould be dipped for some time in water, and afterwards dried, before it was ufed.

29. Thus having cleared up my conjectures, as much as they could be, by thefe preliminary experiments, and got fome infight into the proportions of the different parts of the machine, I proceeded to its conftruction, and finifhed it in the following manner:

Defcription of an HYGROMETER.

30. Tab. XVIII. figure 1. fhews the fection of the interior part of the inftrument, of its true length, in the direction of its axis.

The firft part to be defcribed, being in fome meafure the foul of the hygrometer, is an ivory tube, *a a b*, open at the end *a a*, and clofed at *b*. It is made of a piece of ivory, taken at the diftance of some inches from the top of a pretty large elephant's tooth, and likewife at the fame diftance from its furface, and from the canal which reaches to that point (68). This piece is to be bored exactly in the direction of its fibres; this hole is to be very ftreight, and its dimenfions are $2\frac{1}{4}$ lines in diameter, and two inches 8 lines in depth from *a a* to *c*.

31. Prepare

31. Prepare after this a brafs cylinder, about $3\frac{1}{4}$ inches long, and to one of its extremities fix the pully proper to receive the ftring of the bow when the piece is turning. This cannot be done too carefully, both to make it perfectly round, and to fit it exactly to the hole of the ivory tube; its extremity muft even be rounded, that it may be applied clofely to the bottom of the hole. Having then roughly prepared the outfide of the ivory tube, and introduced into it the brafs cylinder, put both pieces thus united upon the turning wheel, and find out on the outfide bottom of the ivory tube, the point which anfwers to the axis of the brafs piece, in order that this may turn exactly upon its axis. It is with this view that the brafs cylinder is made longer than the ivory tube.

32. All thefe precautions are defigned to make the fides of this tube of an equal thicknefs, viz. $\frac{3}{16}$ of a line, except at the two extremities. At the bottom b the tube ends in a point, and at the top $a\,a$, it muft for about two lines be left a little thicker, in order to enable it to bear the preffure of another piece, which is to be put into it. Thus the thin or hygrometrical part of the tube will be reduced to $2\frac{1}{2}$ French inches, including the concavity of the bottom.

33. Before this piece is ufed, put it into water fo as that the external part alone be wetted by it, and leave it there till the infide be every where covered with the dew I mentioned before (25). This will take place in a few hours; I have given the reafons for this operation (28).

34. The glafs tube intended for this hygrometer muft be about 14 inches long. Its lower end is feen

in

in *d d e e* (fig. 1.). Its internal diameter is about $\frac{1}{7}$ of a line. The reason why it should not be sensibly less will be given hereafter (52); and if it was sensibly larger, the variations of height in the mercurial column would not be considerable enough. On the dimensions that I propose, when the hygrometer is put into melting ice, in a fine summer day, the mercury falls about six inches in the tube. The outside diameter of this tube should be about two lines, in order that the part *g g* of a brass piece through which it passes, and which is to enter into the ivory pipe, be as thin as possible.

35. The·glass tube, as I said before, should have belonged to a thermometer. Its extremity widens of course towards the ball; which will be of use, when the mercury is poured into the hygrometer, in order that it may drive the air before it, by rising from the ivory pipe into the glass tube. To preserve this widening, break the ball of the thermometer by striking against the bottom; and with pinchers take off the rest little by little, and make the extremity cylindrical by grinding it upon the wheel. The same must be done at the top, which I suppose to have been made to end in an olive or small reservoir for the filling of the thermometer. This widening is likewise to be saved for the reasons hereafter to be mentioned (52).

36. The piece *f f g g* is intended to join the ivory with the glass tube. It is of brass, shaped as in the figure. A cylindrical hole is bored through it, which holds the glass tube as tight as possible, without danger of breaking it; and its lower part is to enter with some degree of difficulty into the ivory pipe.

37. To

37. To hinder the part of that tube, which en-closes the brass piece, from being affected by the variations of the humor, which might sometimes prevent a sufficient pressure, I cover this part of the tube with a brass verrel, represented in *b b i i*. It must enter with force, and will henceforth be con-sidered as part of the ivory pipe.

38. To unite those pieces together, I make use of gum lac, or of mastich, which melts by the heating of the glass and the brass. I first cement the brass piece with the glass tube by introducing the tube, and leaving it at first at an inch distance from the place where it is to be fixed ; I then hold this end of the tube over live coals, by bringing it nearer and nearer, and turning it, that both that and the brass piece be every where equally heated ; and when they are hot enough to melt the gum lac, I rub the glass tube with it, and push the brass piece to its place by means of a hollow bit of wood, drawn beforehand over the tube for this purpose. As the brass piece advances, the lac accumulates towards the end of the tube; I take away the superfluous part, but leave a slight coat of it over the end of the brass piece, in order to preserve it from the contact of the mercury that might corrode it. When this piece is properly placed, and still warm, I cover with lac its cylindrical outside, and introduce it into the ivory tube, which has been somewhat warmed by holding it near the fire, in order that the lac may stick more closely to it. As soon as these pieces are cold, they are found very strongly cemented together, and neither mercury nor water can make their way between them.

3   39. The

39. The introduction of the mercury is the next operation. I first roll a slip of paper three inches wide over the glass tube, and tie it fast to the extremity which is nearest to the ivory pipe. I then introduce into the tube a horse hair long enough to enter the cylinder by one end, and to have the other rise three or four inches beyond the orifice of the tube. I then raise the paper which has been shaped round the tube, and use it as a funnel to pour the mercury into the instrument, which I hold upright. The purest quicksilver ought to be employed for that purpose, and it will therefore be proper that it should be revivified from cinnabar. I poured it then into the paper funnel, from whence it easily runs into the tube, with the assistance of some gentle shakes. The air which it drives before it comes out along the horse hair. Fresh mercury must from time to time be supplied, to prevent the entire emptying of the paper tube, and the running in of the mercurial pellicle, which the contact of air always produces upon the surface.

40. Some air bubbles generally remain in the tube; they may be seen through the ivory pipe, which is thin enough to have some transparency. These being collected together by shaking, must be brought to the top of the tube, and expelled, by means of the horse hair. To facilitate this operation, some part of the mercury must be taken out of the tube, in order that the air may be less obstructed in getting out, and the horse hair have a freer motion to assist it.

41. Air, however, cannot be entirely driven out in this manner. It is the weight of the mercury,

with which the tube is for that reafon to be filled,
that in time completes its expulfion, by making it
pafs through the pores of the ivory. To haften this,
I place my hygrometers in a box made on purpofe;
and this I fix pretty nearly in a vertical direction, to
the faddle of a horfe, which is fet a trotting for a few
hours. The fhakes fometimes divide the column of
mercury in the glafs tube, but it is eafily reunited with
the horfe hair. When, upon fhaking the hygrometer
vertically, no fmall tremulous motion is any longer
perceived in the upper part of the column, one may
be fure that all the air is gone out.

42. I now come to the operations requifite to
make the fcale of the hygrometer, and firft of all to
that which determines the bafe (15). This may be
done as foon as the air is gone out. I then fufpend
the inftrument in a veffel filled with ice mixed with
the water it produces in melting. I take care to
fupply the melting by recruits of frefh ice, during
the courfe of this procefs, which lafts ten or twelve
hours. In the firft hour, the mercury finks above
one third of the fpace it has to go through; it advances lefs in the fecond; and its motion leffens thus
gradually, till it appears ftationary, which frequently
happens after feven or eight hours, and it remains
two or three hours in that fituation. The ivory being then become more tranfparent on account of humidity, a very thin dew is perceived by a certain play
of the light on the furface of the quickfilver. Laftly,
the mercury begins to reafcend; the operation is ter-
minated; and fmall drops of water, as I expected,
are at that inftant feen upon its furface (26).

43. 1

43. I follow the laſt ſteps of the mercury in its fall, by means of a fine ſilken thread fixed very tight around the tube. This is left at the loweſt point it has been brought to. If this point be too low, relatively to the frame of the hygrometer, freſh mercury is poured in, and the thread proportionally drawn up higher; if too high, I take off ſome of the mercury and lower the thread; and in both caſes make uſe of the horſe hair. This muſt be done when the mercury ceaſes to fall, in order that the place where the thread is to remain may be immediately determined by this operation.

44. This point thus fixed is named o in my hygrometer; it is that in which dryneſs is nothing (if I may be allowed to expreſs myſelf ſo), ſince it is that of extreme humidity, in a given heat; viz. that of melting ice. From this point are reckoned all the degrees I am now going to ſpeak of; which thus become degrees of exſiccation.

45. The laſt eſſential operation is that by which the ſize of the hygrometrical degrees are determined; and this I ſhall deſcribe by an example. It muſt be remembered that the hygrometer's tube was originally a thermometer (23). I take it in this firſt ſtate, in the inſtance I am going to give. The diſtance between the thermometrical points of melting ice and boiling water, at twenty-ſeven French inches of the barometer, was found to be 1937 parts of a certain ſcale. I broke the bulb of this preparatory thermometer, in a baſon, in order to receive carefully all the mercury that it contained. This being weighed in nice ſcales, amounted to 2 *on.* 11 *dr.*

12 *gr.* or 1428 grains. All the pieces of my hygrometer being put together, it weighed 373 grains,
and when filled with the proper quantity of mercury
833. It confequently contained 460 grains of
mercury.

46. By the rule above given (23), the extent of
the hygrometer's degrees, ought to be to that of the
degrees in the preparatory thermometer, in proportion of the refpective weights of mercury in the hygrometer and thermometer ; and confequently as the
weight of the mercury in the thermometer is to the
weight of the mercury in the hygrometer, fo is any
given interval in the thermometrical fcale, to the
correfponding interval in the fcale of the hygrometer. Confequently in our example as 1428 : 460
:: 1937 : 624 (nearly); and the correfponding in-
tervals on the fcales of the thermometer and the hygrometer, ought to follow the proportion of 1937
to 624.

47. I call the diftance between the two fixed
points of heat in the thermometer the *fundamental
interval*; and I fhall call the *fundamental line* in
the hygrometer that of which the length corre-
fponds to this interval. Thus the *fundamental interval* in the preparatory thermometer, being 1937
parts of a certain fcale, the *fundamental line* of my
hygrometer confifted of 624 parts of the fame
fcale. This example may fo eafily be applied, that
it will be unneceffary to dwell any longer upon this
fubject.

48. Having thus got a *fundamental line* in the
hygrometer, I had it in my power to divide it into
as many parts as I thought proper : my choice was

naturally

naturally to be determined by the fimplicity of a proportion between the degrees of the thermometer, and
thofe of the hygrometer, becaufe this laft was to be
corrected by the firft, on account of the effects of
heat (24). My firft thought was to divide this *line*
into 80 parts, agreeably to the divifions of the fundamental interval in what I call the *common ther-*
mometer in my book upon the air, which I fhall always be underftood to mean in this paper. But as
the minutenefs of thefe degrees was found to be in-
convenient and fuperfluous, I determined to make
them double, by putting only 40 in the length of
my hygrometer's fundamental line. It is eafily underftood that thefe degrees, thus fettled, begin to be
reckoned from the place of the thread, which indicates upon the tube of the hygrometer extreme humidity, by the beat o of the *common thermometer*, or
of melting ice.

49. The inftrument with its frame is feen fig. 2.
the dimenfions of which are every way one half of
thofe of the original. It is mounted on deal, that
being the wood, which fuffers the leaft change
in the length of its fibres. The lower part of the
frame is flit through the whole length of the ivory
pipe, in order that the air may circulate freely round
this pipe, and the bulb of a thermometer which I
fhall mention prefently. The hygrometer is faftened
in three parts; viz. at bottom on a fmall bracket,
at top by a tube paffing through a piece either of
hard wood or of metal faftened by fcrews; but
chiefly by means of a brafs wire on the neck of the
brafs piece, which unites the glafs with the ivory
pipe. This piece is laid in a fmall plate of a hard
wood,

wood, which in that place fills a groove originally
made throughout the whole length of the lead-
board.

50. To keep the duſt from getting through the
opening of the tube, I ſhut it up in a ſmall ivory
caſe.   It cannot be ſealed up, becauſe if air was left
in, it would obſtruct the riſing of the mercury; and
if it was exhauſted, the mercury would be puſhed to
the top by the preſſure of the atmoſphere upon the
ivory pipe; as I have experienced it.

51. Hence however ariſes a ſmall inconveniency;
which is, that as the upper part of the column of
mercury communicates with the air, if it remains
long in the ſame part of the tube, or moves but
little in it, ſome dirt may be left on the ſides.   This
I eaſily remedy, by means of a braſs wire, the extremity of which is dentated in the form of a file, in
order to hold ſome bits of cotton, which I put round
it.   The wire is eaſily introduced into the tube, by
means of the widening mentioned before (35).   I
put it in, when the mercury is below the part it has
ſoiled, and eaſily clean it by this means.   It is on
this account that the tubes to be employed are to be
of about ¾ of a line internal diameter.

52. The ſcale of the hygrometer is marked upon
a deal ſlip, which ſlides along the groove I mentioned
before (49).   This, as well as all the other parts of
the frame, muſt be lined with paper, to mark the neceſſary ſcales; and this paper is afterwards varniſhed
over.   Thin plates of ſilvered braſs can be employed
for the ſame uſe.

53. The mobility of the ſcale of the hygrometer
ſerves to correct, in the obſervation itſelf, the effect

I   of

of the heat on the mercury. At the top of this scale is seen an index, over-against another small scale, marked upon the unmoveable part of the frame. The degrees of this small scale are eightieth parts of the fundamental line, and consequently immediately answer to the degrees of the thermometer on the same frame (48). When the index points to o of the small scale, the thread which indicates upon the tube of the hygrometer the point to which the mercury sunk in the melting ice, answers likewise to o in the scale of the hygrometer. This is the case expressed in the figure wherein the thermometer is likewise represented at o of its scale. By first observing the heat therefore, and conducting the index to the point of the small scale, which answers to the actual degree of the thermometer, the hygrometer will only indicate upon its scale the degrees of the humor. For this scale going through the same variations that the heat occasions in the height of the column of mercury, the indications of the hygrometer become just the same as they would be, if the heat always continued that of the point at which extreme humidity was fixed, viz. o. of the common thermometer.

The scale of the hygrometer is carried to the proper point, by means of a knob fixed on a small piece of hard wood or metal, screwed to the bottom of the board, and which affords a free passage to the tube of the hygrometer.

Account.

Account of the firft OBSERVATIONS *made on the going of this* HYGROMETER.

Read June 10, 1773. 54. My firft hygrometer was ready for obfervation at the beginning of laft February (1772), in a rainy feafon. A few hours after it was taken out of melting ice it was already at 54 degrees of its fcale. The next morning it was only at 50, but towards noon it rofe again to 54. I carried it down to my cellar, which being a confiderable depth under ground is commonly very damp. As I went down the ftairs, I perceived that my hygrometer continued falling, fo that when I hung it up in the cellar it was as low as 35.

55. In the evening of the fame day it was at 28½, and the next night at 21½. It continued falling imperceptibly during the reft of this month, throughout the whole of the next, and till the 19th of April. On that day it was at 3½, and confequently very near extreme humidity: but in this interval it had very often rained, and fnowed, and even when the fky was clear over head, the ftreets had always remained wet, fo that it was evident from all the common appearances, that the humidity had gone on confiderably increafing in the cellar:

56. I was impatient to fee the hygrometer rife again in the cellar itfelf, which I could not however expect but with a north wind. At length, on the 20th of April, though the rain ftill continued, the hygrometer rofe half a degree. In the night of the 20th to the 21ft the wind came about to the north, and when I looked at my hygrometer in the morning, I found it at 6½. It continued rifing imperceptibly

ceptibly the whole of that day, and the next morning stood at 9¼.

57. Another circumstance I was anxious to know, was, whether the hygrometer, after having been kept in the cellar so long, would rise, upon being carried up stairs again, to the point from which it had fallen. The importance of this new observation prevented me from pursuing that I had begun in the cellar. I therefore took my hygrometer out of it, and while I was going up the stairs it rose three degrees. This was at six o'clock in the morning. At seven it was already at 17, and at eight at 23¼. From eight to eleven it rose to 43, and at one o'clock stood at 63. After this it fell again, and at half an hour after five was no higher than 50. The sky had been clouded during the last interval.

As the preceding observations relate only to the hygrometer, and not to humidity, I shall confine myself to them. They are sufficient to give an idea of the going of the instrument in the season they were made. I shall hereafter give an account of some observations taken at other seasons.

FIRST EXPERIMENTS *made to discover the Accuracy of this* INSTRUMENT.

58. The most important thing after the preceding observations, was, to try whether the instrument was in reality comparable. To ascertain this, I immediately constructed four more upon the same principles, which were finished on the 23d of August.

59. I could not use my first hygrometer to make comparative observations with the new ones; its tube

being either too narrow or too fhort. The proportion I had fettled between the capacity of this tube,
and that of the ivory pipe, was deduced from the preliminary experiments I had made in the month of
December (29); and had of courfe been found juft,
as long as the fpring lafted. But even before the
new hygrometers were completed, the quickfilver had
rifen in the firft fo as to run out of the top of the
tube. This, joined to fome other previous obfervations, which had convinced me that the diminution
of the humor is much more confiderable on mountains than in plains (76), induced me to fix the dimenfions of the tube of the hygrometer in the manner laid down in the defcription of the inftrument. I
had been in time to follow thefe dimenfions in the
conftruction of my new hygrometers, fo that when
they were brought from extreme humidity to the
ftate of the air in my apartment in the month of Auguft, the quickfilver did not rife too high in them ;
that is, it remained fufficiently below the top of the
tube, to indicate leffer degrees of humidity afterwards.

60. The four new hygrometers have been conftructed with as little reference to each other, as if
they had been made in different countries. By
comparing them therefore, I have been enabled to
judge of what might be expected from the agree-
ment of inftruments of this kind. This is what I
have found.

When I have obferved them in places where it
appeared likely that the humor would be equally dif-
tributed among them, the utmoft of their difference
has been ufually from 19 to 21. Their greateft
height, for inftance, in my room with the windows
fhut,

fhut, has hitherto been 94, 99$\frac{1}{2}$, 100$\frac{1}{2}$, 105$\frac{1}{2}$, in the fame moment; which is pretty nearly in the proportion of 19 to 21, between the hygrometer which remains at the loweft, and that which is at the higheft.

61. Befides this difference between the relative altitudes of thefe four inftruments, I have obferved another kind of irregularity in them, which is, that they do not always preferve the fame proportion to each other. Thefe variations are undoubtedly in part owing to the caufe itfelf of their motions; that is, to the unequal diftribution of humor even in places very near each other ; but I have reafon to afcribe part of them to fome defect in the inftruments themfelves. I fhall hereafter return to thefe caufes, and give them a clofer examination.

Considerations *on the Degree of* Accuracy *that has been obferved.*

62. Notwithftanding the defects I have mentioned were rather evident, I was not diffatisfied with this firft trial. I never imagined that I had forefeen every thing, and confequently could not expect to arrive at a fufficient degree of exactnefs without the help of experience ; the irregularities therefore which appeared in the execution, did not make me defpair of being able to perfect that inftrument.

63. My hopes in regard to this were at firft only grounded upon general reflections. I recollected what the barometer and thermometer had been when they firft came out of the hands of their inventors; and obferved that in fome refpects they were more

irregular

irregular than my hygrometer is at prefent. Though
the firft of thefe inftruments was very fimple in itfelf, yet barometers hung up in the fame places ufed
to vary three or four lines from each other. Some
of the members of the French Academy have been
themfelves engaged in confidering a barometer that
always kept 18 lines below the reft, and they have
formed various hypothefes to account for this difference. The variations therefore of the barometer,
though obferved only in the fame place, were much
greater than thofe between my hygrometers.

64. Nor was the thermometer itfelf, which is
now brought to fuch a degree of accuracy, much fuperior at firft to our hygrofcopes, for the purpofe of
comparative obfervations. The firft philofophers who
treated of it knew nothing of any fixed point or determined degree in it; they knew nothing even of
the effect produced by the difference of liquids.   In
this ftate of uncertainty the Royal Society adopted the
moft prudent plan that could be thought of; by
giving its fanction to a thermometer to ferve as a
ftandard for the conftruction of thofe which philofophers fhould make ufe of. After this fome men
of genius endeavoured to eftablifh fixed principles for
the making of this inftrument.   Sir Ifaac Newton led
the way, but the utility of his firft attempts was not
fufficiently attended to.   Fahrenheit and Reaumur
then laboured with great care to fettle this point, and
we are much indebted to their inquiries.   But Fahrenheit's principles were foon rejected, as being too
uncertain, though his fcale was preferved; and Mr.
de Reaumur's, though in appearance admitted for a
longer continuance, were in fact fo indeterminate,

2   that,

that, without perceiving it, a deviation from 80 to
104, was made in the space between the two fundamental points of his thermometer.

65. If in the same manner we trace the origin of
all instruments designed for nice mensuration, we
should find that have all been defective at first, and
gradually brought to perfection, when men of genius
have thought them worthy of their attention. Thus
from the first watch, which depended entirely upon
the unequal and uncertain action of a spring, a suc-
cession of attempts has produced Mr. Harrison's
valuable time-keeper; and from the first balances,
which were either too heavy, or too light, we have
attained to those scales of Mr. Matthey * as easily
turned as they are accurate. What however is
still more astonishing is, that, notwithstanding the
importance of having fixt measures for the dimensions of bodies, we have not as yet used any in
practice, but such as must be modeled immediately
from others.

It is true, that in the construction of an Hygrometer, I was assisted by the general notions of regularity,
gathered from the construction of other measures
of the same kind; and of course my hygrometer
is much forwarder in this respect than the thermometer, for instance, was in its origin. I therefore only compare the difficulties peculiar to the hy-

---

* An excellent mechanic, whose death is a loss to a king
who knows the value of men of merit. He was a native of
Vale-Orbe, in the Pays de Vaud, and in the service of his Sardi-
nian Majesty; and has written a *Treatise on Balances*, which
serves as a law to all the scale makers in the dominions of that
prince.

grometer,

grometer, to thofe that firft occurred in all meafures
of phyfical caufes ; and I think that as the latter have
been furmounted, we fhould not defpair of con-
quering the former. In a word, it is certain that all
our inventions only approach towards perfection by
degrees, without ever attaining to it entirely; and
for this very reafon, we have a right to expect they
will always be drawing nearer and nearer towards
it.

Upon thefe notions chiefly, I have raifed my
hopes, either that my hygrometer will in time become more perfect, or at leaft that it may excite
new ideas, which, will at length, though perhaps by
fome other road, lead us to a true meafure of the
humor. As the hope of attaining an end, is one
of the moft powerful affiftants towards really arriving at it, I flatter myfelf at leaft that I fhall have
given birth to a reafonable one upon this fubject.

Firft Views to improve the HYGROMETER.

66. The idea I entertain, that it is neceffary a
number of attentive men fhould concur, to improve
the human inventions ; has induced me firft to mention the general reafons I had, for hoping that the
hygrometer would be perfected. I fhall now proceed to give fome particular reafons on which this
hope is founded, and which are collected from the
remarks I have already made upon my inftrument,
during the little time I have had to obferve it.

The firft, and one of the moft important of thefe
remarks, is, that the ivory pipe belonging to that
hygrometer which is always the higheft upon its
fcale,

scale, happens at the same time to be the thinnest of them all. What connexion there may be between these two circumstances, must be determined by experience. But in the mean time it appears to me that if the fibres of the ivory are interwoven with each other; they will make so much the less resistance either to the being separated or brought closer to each other, in proportion as the bundles of these fibres have a less degree of thickness. Whether this remark is of consequence, or not, we shall at least run no risque in making these ivory pipes always exactly of the same thickness. This indeed was my intention in those which I have made, but unfortunately I thought I should have been able to turn them upon cylinders of hard wood; and found too late, that no accuracy could be expected from this method. It was to remedy this inconvenience, that, in speaking of the manner of turning this piece, I have recommended brass cylinders (31).

67. The same precaution is likewise necessary to be taken, that we may be certain of giving to every pipe an equal degree of thickness throughout the whole of its circumference: a circumstance no less essential than the former; since I have observed in those of my hygrometers whose pipes have not an equal thickness, that they bend, more or less according to the degree of humor to which they are exposed.

This is probably the principal cause why these instruments do not always preserve the same proportions to each other (58). For the pipes not bending according to the same law, there must be an irregular

regular change in their capacity, and confequently
in the height of the mercury in the tubes. The
differences of this kind which I have had occafion
to obferve, are not indeed very confiderable; but,
however trifling the caufe of an imperfection may
be, it is ftill ufeful to remove it; were it only to
affift us in difcovering caufes of greater imperfections,
by making their effects the more evident.

68. But to make the ivory pipes keep ftraight,
we muft attend to a circumftance ftill more important; which is, that the texture of the ivory be the
fame in the whole circumference of the pipe. There
is a fenfible difference in the organization of the external, middle, and internal parts of the fame elephant's tooth: nor is it impoffible that, befides this
difference in the nature, and vifible arrangement of
its fibres, there may be another arifing from their
degree of tenfion; fo that fome fibres may be more
difpofed than others to relaxation, after the tooth
has been cut to pieces. Suppofe then that any of thefe
differences fhould exift in a pipe, that is, if one of its
fides fhould be more porous, or of a weaker texture,
than the other; or if its fibres fhould be more difpofed
to relaxation; this pipe will take a bend, either for
a conftancy, or for a time; and the hygrometers in
which it is ufed, will not of courfe agree with the
others. We muft therefore endeavour to make thefe
pipes with a part of the tooth that is homogeneous:
that which I believe to be moft fo, within a certain
extent, and which for that reafon I have advifed, is,
the part which is between the center and the furface,
and at fome inches diftance from the apex of the
tooth (30).

69. There

69. There is another reason why this different organization of the different parts of the elephant's tooth makes it necessary to determine exactly the parts that are to be made use of in hygrometers. Without this precaution it might happen that the pipes, which ought to be similar in every respect, might be made of substances that really differed in their dilatability and sensibility; that is, of substances which the humor might affect more or less strongly, or more or less quickly. This consideration will perhaps oblige us to determine both the size of the tooth, and the distance at which the piece ought to be cut off from its apex: for the organization may with equal probability vary in teeth of different thickness, and from the apex to the base; as it does in the breadth of the same tooth. I was not sufficiently certain of the success of my instrument, to take all these precautions when I first set about it, but at present I believe them to be important.

70. There is still another precaution, which indeed I thought necessary from the beginning, but which I could not manage as I wished for want of proper tools; that is, to perforate the ivory pipe exactly in the direction of its fibres. For let the channel have ever so small a degree of obliquity with respect to this direction of the fibres, these fibres will necessarily be cut in different places; which weakening the pipe where it happens, neither its dilatations nor its contractions can of course be regular.

71. I own here are a great number of precautions; but they will not surprize true philosophers. They are accustomed to observe the operations of

nature clofely ; and know that the regularity of her
proceedings is connected with a forefight whch is
limited to us, by nothing but the limits of our abi-
lities in tracing it ; and confequently, when art attempts to imitate nature, it can only fucceed in as
much as it is attentive to imitate her care.

72. I believe that the hygrometer may farther
acquire a perfection of the fame kind as that which,
in conformity to an idea of my worthy friend Mr.
Le Sage, I have given to the thermometer ; that is,
that we may make its degrees correfpond with equal
differences in the humor; as I have made thofe of
the thermometer correfpond with equal degrees of
heat. The way in which I think this might be done,
would be to fufpend near one of the hygrometers, in
a proper veffel which fhould be placed in one of the
fcales of a balance that turns very eafily, fome fubftance remarkably greedy of the humor ; the aug-
mentations or diminutions of weight in which fubftance, might be compared with the going of the
hygrometer, firft in the fame, and afterwards in different degrees of heat. I hope too, that by a fre-
quent repetition of thefe obfervations, at times when
the variations of the humor are more or lefs fudden,
we fhall at laft fucceed in correcting the errors that
may attend them, from the lofs of its own matter
the fubftance made ufe of may probably fuffer by
evaporation.

73. Thefe are not the only remarks I have made
upon my inftrument, but I did not care to mention
any but fuch as have appeared to me moft certain.
The others are uncertain, and require longer obfervations. I fhall only add therefore, that it will ftill

5   be

be neceffary to make frefh experiments, in order to determine the length of time that the ivory pipes ought to remain in water, and how long they muft afterwards be expofed to the viciffitudes of the air (28), or in general to what preparation they muft be fub-mitted, in order to acquire a lafting degree of con-fiftence before they are made ufe of. For this pur-pofe it will alfo be expedient to compare hygrome-ters recently made, with older ones, both to afcer-tain whether they have undergone any alterations, and in what degree. I likewife am of opinion, that when we wifh to fix the point of extreme humi-dity, we muft be very careful not to make ufe of any ice but what is very clean, as well internally as externally; left any duft fhould ftick to the ivory pipes, which might hinder the water from pene-trating into the pores : this is what I thought of my-felf too late. I do not know whether for the fame reafon it would not be right to wafh thefe tubes with fpirits of wine before we put them into the water, to remove any greafy fcurf they may have gathered by handling; and afterwards to repeat this at times, in order to carry off any little depofit of various kinds, which may in courfe of time have been left upon them by the air. Moreover it will be right to inquire whether there is not a difference between the effects of the heat upon the ivory of the hygrometers, and upon the glafs of the ther-mometer, fenfible enough to be attended to, in cor-recting the effects of this caufe upon the hygrome-ter.

74. Having already difcovered fo many caufes, more or lefs probable, of the differences I obferved

in

in my hygrometers, I think it reasonable to hope
that this instrument will receive a sensible degree of
perfection on a second trial; and that in time it will
be brought to a sufficient degree of accuracy. It is
true there are some difficulties in the way of this:
but have we not sufficient motives for endeavouring
to overcome them? The air we breathe, and that
which surrounds us; the places we inhabit, and
those which serve either to enclose or to preserve
so many different bodies intended for our several occasions, are all of them more or less filled with that
substance, differently modified, to which I have given
the name of *humor*. It also produces very sensible
effects in them; some of which very properly excite
our curiosity, others may be turned to our advantage, and many of them essentially affect our health.
It is therefore of great consequence to natural philo-
sophy in general, and to œconomy, and medicine, in
particular, that we should obtain a measure by which
we may, with some degree of certainty, estimate the
local and actual qualities of this substance, and by
this means forosee its effects; which for the gene-
rality we only become acquainted with after they are
produced. These sciences are not perhaps less con-
cerned that we should discover the nature itself of
this agent, and the different manners in which it ope-
rates: the knowledge of which may enable us to
avail ourselves of reason in the investigation of certain effects, which, without such helps, might escape
our observation. As these are the several uses of an
exact hygrometer, we may readily perceive how
many new tracks such an instrument may open to
us, in our investigations of nature, which however

2   we

we fhall not owe to one man alone, but to the joint
labours of feveral.

Account of fome of the firſt Phænomena of the
Humor obſerved with the Hygrometer.

75. Though my firſt advances in this new track
of obſervation are as yet very uncertain, yet I will
not omit giving fome account of them. They will
at leaſt ferve to give fome idea of the going of the
inſtrument, as well as of the nature of the agent by
which it is governed.

The firſt obſervation I attempted of this kind was
with a view to one of the objects which made me
deſirous of having an hygrometer. Theſe objects
are all comprized in a general fyſtem concerning
vapours, which I have given in my work upon the
Modifications of the Atmoſphere. I fhall therefore
only mention here one particular confequence of that
fyſtem, which it was my immediate point to verify;
namely, that a certain augmentation of heat, we
always perceive at every feaſon upon the approach
of rain, is owing to a more than ordinary quantity
of vapour; and that, on the contrary, it is to their
diminiſhed quantity that the leſſer heat of the upper
parts of the atmoſphere is in great meaſure to be
afcribed.

76. This latter confequence was fupported by an
accidental obfervation I made in September, 1770,
upon a mountain of the Faucigny, at the height
of 1560 toiſes above the level of the fea. An iron
ferule, which ferved to unite the ends of a cleft
ſtick, and which had been fixed on the ſtick with a
hammer

hammer upon the plain, in fine weather, came off
of itſelf upon the top of the mountain. When this
happened, the thermometer, which I called common,
though expoſed to the ſun, was only three degrees
above 0; while on the plain it was at 18 in the
ſhade. This phænomenon, joined with ſeveral others
I obſerved at the ſame time, confirmed me in my
opinion, that one of the reaſons why the upper parts
of the atmoſphere have leſs heat than the lower, is,
that they contain leſs humor.

77: With this notion, it became a very intereſt-
ing inquiry to know the different degrees of humidity habitual to the different heights of the atmo-
phere. Of courſe therefore this was the firſt obſervation I thought of, as ſoon as I had added an hygrometer to the other inſtruments contained within
the box of my portable barometer. I undertook
therefore to aſcend Buet (the name of that high
mountain) a ſecond time. My companions in this
expedition were Mr. Dentan, a very intelligent young
philoſopher, and my brother, who, having aſſiſted
me in all undertakings·of any difficulty, had been a
witneſs of that fact which was the object of my pre-
ſent reſearches.

78. At our ſetting out, on the 29th of laſt Auguſt,
the hygrometer was at 86 in my apartment, and the
barometer at 27 p. 1 line. We were in hopes of
fine weather, becauſe it is generally fair in this country when the barometer at Geneva is above 27
French inches. Soon after we ſet out, we began to
perceive that the power of the ſun was greater than
might have been expected for the ſeaſon. From this
circumſtance I concluded that the barometer muſt
fall;

fall, and in fact we found it lower at every place in our way, where we had before obferved it in fine weather. The fky, notwithftandihg, was ftill clear, and continued fo the next day, when we began to afcend the mountain, about two o'clock in the afternoon, in order to pafs the night in the higheft cottages, that we might have more time to gain the fummit the next day.

79. Before we left Sixt (an abbey at the foot of the mountain), I expofed the hygrometer in open air, and in the fhade it ftood at 94. The thermometer at the fame time was at 19 in the fhade, and at 24 in the fun. At five o'clock we reached a place above 300 toifes above the abbey; commanded on all fides by mountains, and on that account called *Les Fonds* (or The Bottoms). Here we obferved the thermometer and hygrometer. The former, when expofed to the fun, ftood at 15¼, and the latter rofe to 96 in the fhade. We obferved them again in the fame manner about half an hour after fix, in a place that was pretty open, and higher by 160 toifes than the former. The thermometer ftood at 15, and the hygrometer at 106. It wanted but a quarter of nine, when we came to the cottages where we were to pafs the night; though they were not above 30 toifes higher than the place we ftopped at laft. The higher we went, the clearer the fky appeared; in fo much that, notwithftanding the ufual augmentation of humor in the air after funfet, when the fky is not clouded, upon expofing the inftrument to the air, about ¼ after ten at night, we found the hygrometer at 123, and the thermometer at 13¾. They both fell in the night, and on our
    fetting

Setting out the next morning, the former had got down to 109, and the latter to 12.

80. In the two laſt mentioned obſervations the hygrometer had been expoſed long enough to the open air, to conform itſelf to the degree of humor prevalent in the place; but we had not time for the obſervations I was moſt deſirous to make with accuracy. The hygrometer being uſually ſhut up in the box of my barometer, it would have been neceſſary to have left that open ſome time, in order that it ſhould adapt itſelf to the ſtate of the air, and we could allow but a very ſhort time for theſe obſervations.

81. The firſt of them was made at nine in the morning, at the height of about 1000 toiſes above the plain. The ſky appeared clear over head, but the plain was darkened with vapours. The thermometer in the ſun ſtood at 13¼, and the hygrometer roſe to 115 in the ſhade.

82. It was two in the afternoon when we reached the top of the mountain, which is always covered with an enormous maſs of ice and ſnow. We found there a very ſtrong ſouth wind, which is the warmeſt wind in our plains: beſides this, we were nearly at the hotteſt time of the day: and yet the thermometer, upon being expoſed to the ſun, ſhewed only 6. The wind, and the coldneſs of this region, obliged us to quit the ſummit in a quarter of an hour, during which the hygrometer had riſen only to 119; but we judged that it was not yet ſtationary.

83. In this ſhort time we experienced a new effect of the diminiſhed humidity of the air, which ſurprized us all three very much. We found our ſkin
withered

withered and pale, so that both to the sight and to the touch, it resembled much a dry and shriveled bladder. Notwithstanding this we were sensible of no other inconvenience but what arose from the wind and the cold: the action of the lungs and the functions of all the other parts of the body were perfectly free, though the barometer was only at 19 inches, 6 lines and a half.

84. We quitted the summit at about a quarter after two, to shelter ourselves from the wind behind some rocks, which were nearly 50 toises lower. Here we stayed about an hour. During this time the hygrometer, exposed to the air, but always in the shade, rose by imperceptible degrees to $132\frac{1}{4}$. It would probably have risen higher, had not we been obliged to quit this place, where the clouds began to gather, in order to reach the cottages before night. It was indeed already too late before we thought of retiring; for we were overtaken by the night, and a thunder storm, at a sufficient distance from our hut to expose us to the greatest danger of being lost, notwithstanding our guides, but for the assistance of two women, whose humanity deserves the highest commendations. These women, who lived in our cottages, being apprized of our distress by our cries, notwithstanding the storm, and the scarcity of wood in these places, came out to kindle a great fire at the foot of the rocks on which we were wandering amidst the precipices, in total darkness; and sometimes with great difficulty keeping the fire alive, sometimes advancing towards us with fire-brands till the wind and rain extinguished them, and endeavoured, with the most unaffected concern, to point out to us the

VOL. LXIII.   M m m   path

path we ought to keep. At laſt, animated by the courage of theſe women, directed partly by their light and partly by their cries, we at length reached the cottage, much more affected with the humanity of theſe good people, than hurt with the dangers and fatigues we had undergone.

85. The ſtorm laſted a great part of the night, and it rained almoſt without intermiſſion. Notwith-ſtanding this, the hygrometer, when expoſed to the air the next morning, ſtood at 105, and the thermo-meter at 10. As we were uncertain how long the rain would continue, we ſet out at eight in the morning on our way down. The rain hardly ceaſed the whole morning, and was ſometimes accompanied with hail; it ſtill continued raining when we arrived at the abbey about noon, notwithſtanding the hy-grometer ſtood there at 99, that is to ſay, five degrees higher than when we ſet out; but the barometer, which had fallen the two preceding days, was now beginning to riſe; the thermometer was at 14.

86. We learnt at Sixt, that, at the very time we were driven from the ſummit of the mountain by the diſagreeable coldneſs of the air, they had felt an ex-ceſſive degree of heat, and likewiſe that the ſtorm had been very violent in the night. This ſtorm, as we found two days after at Geneva, had extended it-ſelf all over the plain. We found likewiſe, from the obſervations that had been made there in our ab-ſence, that a thermometer expoſed to the north, con-ſequently out of the ſun, had been at 23½, at the very time that ours, at the top of the mountain and in the ſun, had been only at 6.

87. As,

87. As, in mentioning the particular purport of the foregoing obfervations, I have not explained my fyftem concerning vapours, I fhall not here ftop to draw the confequences that may be deduced from them in favour of this fyftem. Indeed, to fay the truth, I·think them too few and too imperfect to conclude any thing from them as yet. I have only related them, as I declared at firft, to give a general notion, both of the going of my hygrometer, and of the inquiries that may be purfued with its affiftance. It is with the fame intention that I proceed to relate fome obfervations of another kind.

88. Some accidental obfervations had made me fufpect that the immediate action of the fun upon my hygrometer produced a drying, which might not be wholly occafioned by the real ftate of the air with refpect to the humor, but might depend in fome meafure upon fome fingular property of the folar rays, which we fee produce effects upon fome bodies, not immediately to be accounted for by the ordinary laws of heat. This firft remark induced me, as I have taken care to mention, always to obferve the hygrometer in the fhade upon the mountains of Sixt. At my return, I determined to examine more accurately whether my conjecture in this refpect had any foundation.

89. The firft thought that occurred to me for this purpofe was, to obferve two hygrometers at the fame time, one in the fhade and the other in the fun, very near each other, that the fame air might circulate freely round them. The air of the country having appeared to me more proper for this obfervation than

M m m 2   that

that of the town, I determined to obferve at the fame
time the variations of the humor in the open air for
a whole day together. There are doubtlefs many
varieties in this refpect; nor indeed fhall I determine
from this obfervation, any thing more than the ftate
of the open air during one day, and in one particular fpot.

90. I made my obfervation the 13th of September 1772, in a garden fituated to the weft of our
lake, and only feparated from it by another garden
and fome buildings. There I hung up a couple of
hygrometers which I kept perfectly infulated, one of
which had no other frame but a fcale fixt to its tube,
and the other was in a frame whofe opening at the
height of the ivory tube was of a confiderable fize.
They were four feet and a half above the ground,
and at the diftance of a foot from each other. A
piece of pafteboard about 12 inches in breadth,
placed at a foot's diftance from the hygrometer which
was not mounted, was intended to fhelter it from
the fun. Each hygrometer had a thermometer clofe
to it, the ball of which was not in contact with any
thing. I have proved in my work, that it is neceffary to keep this ball infulated, in order to obferve
the heat of the free air.

91. One of the hygrometers I made ufe of for
this obfervation, was at 93, and the other at 96½
in my room the night before. In order to correct
this difference, which I fhould fuppofe to be in proportion to their height, I would always add about
$\frac{1}{17}$ to the height of that one which kept itfelf the
loweft, that there might be no difference between
them, but fuch as fhould be produced by the difference

ference of the quantity of the action of the humor.
The hygrometer which remained at the lowest was
the one that was always in the shade, and was not
mounted. It was the same upon which I had made
my observations in the mountains of Sixt. I suf-
pended them both in the garden I have been speaking of, about 6 in the morning; the plants were
covered with dew; the sun, being just rising, could
not yet shine on the garden. As soon as the hygrometers were exposed in the open air, they both fell
very rapidly, but the one which was without the frame
fell much faster than the other. They both were
continuing to fall, when the sun began to shine in
the garden. The following is an account of their
progress, and of that of the thermometers during 19
hours. The action of the heat upon the mercury
of the hygrometer is corrected upon each of them,
from the observation of the thermometer joined to it,
so that there only remains that of the humor.

A Table

A TABLE of OBSERVATIONS made on the 13th of September, on two HYGROMETERS, the one in the Shade, and the other in the Sun, each of them accompanied with a THERMOMETER.

	Hour of the Day	Therm. in the Shade.	Hygrom. in the Shade.	Hygrom. in the Sun.	Ther. in the Sun.
The Bar. at 27 Fr. inch. 1 line. The fun did not fhine yet on that part of the garden.	7	8	29	$36\frac{1}{4}$	8
The fun has now fhone for $\frac{1}{4}$ of an hour on the Hygr. and Therm. which are to ftand expofed to it.	$7\frac{1}{2}$	$11\frac{3}{4}$	$36\frac{1}{2}$	$66\frac{1}{2}$	12
	8	$12\frac{1}{4}$	$43\frac{1}{2}$	82	$12\frac{1}{2}$
	9	13	67	102	$13\frac{1}{2}$
	10	$14\frac{1}{2}$	$76\frac{1}{2}$	109	$15\frac{1}{2}$
	11	15	$87\frac{1}{2}$	116	$16\frac{1}{4}$
	Noon	$15\frac{1}{2}$	$96\frac{1}{2}$	$120\frac{1}{2}$	$17\frac{1}{4}$
	1	$16\frac{3}{4}$	103	126	18
The vapours condenfing in the air weaken the action of the fun.	2	$16\frac{1}{4}$	103	125	$17\frac{1}{4}$
Barom. 27. inch. A South wind begins to blow.	3	$16\frac{3}{4}$	$102\frac{1}{2}$	123	$17\frac{1}{4}$
The clouds rife.	4	$15\frac{3}{4}$	107	133	16
The clouds meet, and the funfhine withdrawn.	5	$13\frac{1}{4}$	$88\frac{1}{2}$	106	$13\frac{1}{4}$
The fun is fet, and the weather quite overcaft.	6	12	$64\frac{1}{2}$	81	12
Barom. 26 inches 11 lines.	7	$11\frac{1}{4}$	50	65	$11\frac{1}{4}$
	8	11	37	50	11
	9	$10\frac{3}{4}$	31	41	$10\frac{3}{4}$
The clouds break, and the dew begins to appear on the plants.	10	$10\frac{1}{2}$	24	35	$10\frac{1}{2}$
	11	10	$20\frac{1}{2}$	$26\frac{1}{2}$	10
The clouds meet again.	Mid.	$10\frac{1}{2}$	$24\frac{1}{2}$	$28\frac{1}{2}$	$10\frac{1}{2}$
	1	$11\frac{1}{4}$	23	27	$11\frac{1}{4}$
It begins to rain.	2	$11\frac{1}{2}$	27	32	$11\frac{1}{2}$

92. The

92. The firſt circumſtance in theſe obſervations that deſerves to be noticed, is the .difference in the ſinking of the two hygrometers when they were expoſed to the air, before the ſun ſhone in the garden. They both of them fell conſiderably, but one of the two 7 degrees and a half leſs than the other. One of the cauſes of this diſparity is probably in the inſtruments themſelves, and is owing to their being differently affected by the action of the humor. There is a difference of the ſame kind obſervable in the thermometers, which are likewiſe more or leſs ſenſible to the impreſſions of the heat even when the bulk of their liquid is the ſame; that is to ſay, they are acted upon more or leſs quickly by the degree of heat which ſurrounds them, according to the thickneſs, or even according to the nature of the glaſs of which the ball is made. Conſequently it is poſſible that the different thickneſs or poroſity of the ivory may have had ſome influence on the going of the hygrometer in this obſervation (66 and 69).

93. But theſe differences in the ivory pipes muſt produce a much greater difference in the ſenſibility of the hygrometers, than thoſe of the glaſs balls can produce in the thermometers; becauſe it is much more difficult for the humor to penetrate the ivory, than for the heat to get through the glaſs. So that any encreaſe of the obſtacles retards the introduction of the humor, much more than that of the heat; and conſequently the difference of ſenſibility muſt be more difficult to be prevented in the hygrometers, than it is in the thermometers.

This ſlowneſs of the humor in pervading the bodies into which it inſinuates itſelf, makes it a de-

5   ſirable

firable circumftance, that the ivory pipe of the hygrometer fhould be the thinneft poffible; in order that it might be more readily affected. This I had forefeen, before I had learnt it from experience; but I was afraid of its being attended with ftill greater inconveniences than that it was intended to remedy; from the action of the mercury againft pipes whofe fides would be thinner. However, this might be tried. In the mean time, I fancy that, for obfervations in which it is abfolutely neceffary that the inftrument fhould eafily be affected, leffer hygrometers might be made, whofe tubes containing a lefs quantity of mercury, would refift the action of it, though with a lefs degree of thicknefs. (Perhaps it would not be impoffible to ufe tubes made of fome very thin quills.) I cannot yet afcertain whether thefe little hygrometers could be graduated by themfelves, or whether they muft be compared with thofe of which I have given the dimenfions; this we fhall learn from experience.

94. The difference there is between the heat and the difcrete humor in the power of diffufing itfelf, occafions in another refpect a confiderable difference in the goings of the thermometer and hygrometer. The heat is brought into a ftate of equilibrium much fooner and with much greater certainty than the humor. Two thermometers accurately conftructed and fixed near each other, in a place where the heat does not change very fuddenly, always agree together. This is not the cafe with two hygrometers: they feldom agree, that is, they feldom preferve the fame conformity to each other, when there is the leaft variation in the humor: at fome times their difference increafes, at others it diminifhes; this can

only

only arife from a difference in the caufe that acts upon them.

95. We may form our ideas of the manner in which the invifible humor diftributes itfelf; from that in which all kinds of vifible vapours are diffufed. We fee them feparate, re-unite, fly off from certain places, rufh into others, and in fhort yield to every impreffion of the air. The motion peculiar to their own particles, which I look upon as the caufe of their elafticity *, is not fufficiently rapid, and the vapours themfelves are too thick to overcome always the contrary motion of the air. This, I believe, is what conftitutes the chief difference between vapours, and the igneous fluid, as far as relates to the power of putting themfelves into a ftate of equilibrium in the air, which is moving. The current of air carried towards a chimney which has fire in it, frees the room from fmoke, and is but a very flight impediment to the diffufion of the heat through it.

96. Though the invifible vapours by reafon of their exceffive thinnefs are more capable of being put in equilibrium in the air than the vifible ones, they are very far from having this property in as great a degree as the heat. Which leads me to think, that part of the difference obferved between my hygrometers, even before funrife, may have been owing to the unequal diftribution of the humor, though the two inftruments were only at the diftance of a foot from each other, without the interpofition of any folid body.

* The fyftem I adopted on that point may be found in my work upon the *Modifications of the atmofphere.*

97. I (hall not attribute intirely to the fame caufe, the great difference obferved between my hygrometers, when one was expofed to the fun, while the other ftood in the fhade. The immediate action of the folar rays, or of the luminous heat, produces a variety of effects, which, as I have faid before, do not appear to follow the fame laws as thofe of dark heat. And if I may be allowed to propofe a conjecture upon this particular point, before fuller experiments have been made, it fhould feem, that the immediate action of the folar rays muft occafion a greater evaporation than what is produced by dark heat, even when they hold the thermometer at the fame height. But let the caufe be as it will, we fee by this experiment, that in a fection of air about a foot wide, through which the folar rays did not immediately pafs, the action of the humor upon the hygrometer was 23 degrees greater than in the place round about; though that of the heat upon the thermometers was only a degree and a quarter lefs; which leads us to conceive how many apparently fmall caufes may contribute to produce fenfible differences in the diftribution of the difcrete humor.

98. Another ufe to be made of thefe obfervations is, to compare them with thofe that I have made in the mountains of Sixt; in order to form a better judgment of the proportion between the different degrees of humidity, in the fuperior and inferior parts of the atmofphere. My hygrometer, held in the fhade upon the fummit of Buet, rofe to 132½, and was not yet ftationary. This is pretty nearly the greateft degree of drynefs obferved in the hygrometer expofed to the fun in the garden; while
the

the one that remained in the fhade, the fame upon
which the obfervation at the mountain had been
made, was not in fact higher than 103, though
marked in the table of obfervations at 107 (91).

99. But the difference between the obfervations
made upon the mountain of Sixt, and thofe I am
fpeaking of, was ftill greater by much after fun-fet.
The 30th of Auguft, at a quarter after ten at night,
I obferved the hygrometer without fide the cottage
upon the mountain, and found it at 123 (79); and
on the 13th of September following, in the plain,
it was not higher than 31 at 9, and 24 at 10 o'clock.
The wind was fouth, and the height of the barometer upon the plain, pretty much the fame during both
the obfervations.

100. It is true that, notwithftanding the fimila-
rity of thefe circumftances, thefe obfervations cannot
be directly compared, on account of the difagree
ment in fome other circumftances. In the firft
place, the difference of fourteen days at this feafon
of the year may have produced a fenfible change in
the ftate of the air. There was already, for inftance,
a confiderable difference in the degrees of the thermometer; it was at 13 and ¼ when the obfervation
was made on the mountain, and no higher than at
10 on the plain. Befides, at this time of night,
there would always be an effential difference between
the upper and lower parts of the atmofphere, even
though in the day time they fhould have the fame
degree of humidity : for the vapours being condenfed
after fun-fet, and thus producing a kind of dew,
they muft necefſarily defcend, and from this very
caufe be more abundant in the low grounds than

N n n 2   on

on the higher ones. I shall add, that though my hygrometer was expofed to the open air on the mountain, as it was in the plain, yet it was not so much infulated there, being tied to the box of my portable barometer. The difference obferved, however, is fo confiderable, that, notwithftanding the concurrence of all thefe particular caufes, I cannot but afcribe it in fome meafure to that general one which I have fufpected, namely, that there is comparatively a lefs degree of humidity in the upper than in the lower parts of the atmofphere.

101. The obfervation of the 13th of September feems likewife to throw fome light upon the phænomena of dew. We know that when the fky is cloudy, there is little or no dew, and it has likewife been obferved from this very circumftance, that the air is not fo much cooled after fun-fet. The caufe of thefe differences appears to me to be, that when there are no clouds in the air at fun-fet, or when they are difperfed, the heat of the inferior air, and that which rifes from the earth, diffipates itfelf into the fuperior regions, and then the vapours which are difperfed throughout the air condenfe and fall down again in dew; but when the clouds are continued, and thus feparate the inferior from the fuperior air, they prevent this diffipation of the heat, and the vapours remain fufpended. And if the fky grows cloudy fome hours after the fetting of the fun, and after the heat has fenfibly diminifhed in the inferior air, it encreafes again in it; becaufe the heat, which continues to rife out of the earth, is accumulated in the inferior air. This appears in the obfervation I am fpeaking of. The clouds having been feparated

5   rated

rated for a while, at 10 o'clock there was fome dew, and the hygrometer fell fenfibly till eleven : but afterwards the clouds clofing again, the heat encreafed, and the humidity evidently diminifhed.

102. I take it for granted here, that the moft common and moft plentiful dew proceeds from the air, and not from the earth, as fome philofophers have imagined. I fhould produce the proofs I have collected of this fact from a multitude of experiments, if it had not been done in an excellent paper, written by Profeffor le Roi, *On the elevation and fufpenfion of water in the air* *. Thefe phænomena of the dew become very interefting examined with the help of the hygrometer, and joined to obfervations of the degrees of faturation of the air with refpect to water, which have been fo ingenioufly imagined, and begun by the author of this memoir. If this part of natural philofophy is ever cleared up, as I hope it will be, we fhall be much indebted for it to the fagacity of this true philofopher.

103. I fhall only mention one more obfervation I have endeavoured to make with my hygrometer, which ought not to be omitted, as it is connected with the principles upon which the inftrument is conftructed. It has likewife a reference to medicine, in as much as one of the objects of that fcience, in its inquiries to preferve our health, is to determine the effects of water at different degrees of heat upon our organs. Ivory being an animal fubftance, the effects produced upon it by water at different degrees

* Mem. de l'Ac. des Sc. de Paris, for the year 1751.

of

of heat, may affift us in difcovering thofe which are produced upon our bodies from the fame caufe.

104. The point o of my hygrometer, as I have before obferved (44), is that of the extreme humidity produced by melting ice. It was therefore of fome importance to know what difference there would be in this point, when the hygrometer fhould be plunged into warmer water. This I endeavoured to find out; and the following is the refult of my firft inquiries.

105. The moment I took one of my hygrometers out of melting ice, I plunged it into water at the heat of 45 degrees of the thermometer that I have called common. It fell fuddenly four of its degrees below the thread which marked its height in the melting ice, but immediately rofe again, and in four minutes reached 8 degrees and a half above the fame thread. Deducting 22½ from the height for the dilatation of the mercury (48), there will remain 14. Confequently the water warmed at 45 degrees of the common thermometer, really made the hygrometer fink 14 degrees below o.

106. Half an hour after this, the water being at 38 degrees, I found the hygrometer no higher than 6¼, that is to fay, $6\frac{1}{4} - \frac{38}{2} = -12\frac{1}{4}$. Confequently the true point of humidity indicated by the hygrometer was 12¼ below o. Laftly, the heat of the water being reduced to 28 degrees the hygrometer was at $3 - \frac{28}{2} = -11$. I was then obliged to put an end to the experiment, which I have not been able to take up again fince, for want of leifure. But

what

what has already been obferved is fufficient to fhew us that the warmer the water is, the more it dilates the ivory (though we faw that the mercury rofe in the hygrometer after having funk for a moment). From hence, I fancy, may be drawn this general confequence, already indeed forefeen, namely, that in *an equal acting quantity*, the warmer the humor is, the more it feparates the particles of thofe bodies which it pervades.

107. I fay, in an *equal acting quantity*; and this is one of the objects which will probably furnifh us with a variety of moft ufeful knowledge, at the fame time that it is moft likely to give the greateft exercife to the genius and attention of natural philofophers. The forementioned experiment proves, that the warmer the water is, the more it dilatates the ivory pipe of the hygrometer, and the fame thing I make no doubt happens with the difcrete humor.

On the other hand, the evaporation being certainly greater in fummer than in winter, there muft of courfe be more vapours in the air in the firft of thefe feafons than in the latter. Thefe then, as it appears, are the two circumftances moft likely to make the hygrometer fall in fummer; a greater degree of humor in the air, and an encreafe of heat. And yet I have already experienced that the mean height of the hygrometer is greater in fummer than in the other feafons. I found my firft hygrometer, which was made in winter, too fhort in the fummer; but it would be of a fufficient length now that we are in autumn. The mean height of the four new ones is already (the beginning of November)

ber) 17 degrees lefs than it was in the months of
Auguft and September.

108. I hope this paradox will be explained, and
that the principles which may clear it up will draw
ufeful confequences along with them. Thofe philofophers who look upon evaporation as a diffolution
of water by air in the manner of menftrua, that
is, by affinity, will eafily apply their principle to the
folution of part of thefe phænomena. The diffolution is greater when the menftruum is warmer, and
confequently the air muft keep a greater quantity of
water in diffolution, and fuffer a lefs part of it to
be precipitated, in fummer than in winter. I cannot but allow that this fyftem is extremely fpecious,
and that many phænomena are very happily explained by means of it. This is what Mr. le Roy
has fhewn us in the memoir I have already quoted;
in which, without contending that air really acts as
a menftruum with refpect to water, he demonftrates,
by a parallel very well kept up, that all the chemical expreffions concerning diffolutions may with
propriety be applied to defcribe the feveral phænomena he examines, relative to the elevation and
fufpenfion of water in air, as well as to its precipi-
tation under different forms.

109. If it was not too common a practice, to
conclude things from words, I fhould in fact think
thefe chemical expreffions very conveniently adapted
to explain a number of thefe phænomena. But I
have rejected them here, on account of this confideration; that when I took in a greater number
of phænomena, I found them no longer accurate,
any more than the general idea of the diffolution

tion of water by air. I have given the reasons for
this in my work, upon the modification of the at-
mosphere; and shall only repeat here, that these modifications of the humor appear to me almost intirely
to be produced by the igneous fluid; and that if the
air has any share in them, it is only as being an
elastic fluid. The particles of those fluids, each according to its degree of power, strike, separate, and
draw along with them those of the humor, and com-
municate to them the elasticity they possess; in the
same manner as they do to the particles of all volatile,
and likewise of all fixt substances which they corrode and decompose.

110. This system will not only furnish a solution
of the paradox which engages our attention, but will, I
believe, carry us much farther. The heat of the
summer keeps the humor in very great agitation;
and though there is more of the humor at this season than in winter, yet this heat will not allow it to
continue either as long a time, or in as great a quantity,
upon the bodies or in their pores. That is the
reason why the hygrometer falls less. But we see at
the same time, that the portion of the humor which
does sojourn, and which I call the active part, has
more power to dilatate the bodies, from the greater
degree of motion impressed upon it by a greater
heat. Consequently the dilatation of the bodies, from
this cause, will be in a compound ratio of the quantity of humor, and of its active force, or of the heat.
And if, for instance, we compare any summer's day,
in which the hygrometer in open air is at the same
degree as on any winter's day, the air on the summer's day will contain more humor than on the

winter's day; but there will be less of it will
act upon the hygrometer; and yet as the active part
will have more strength, the effect upon the whole
will be the same. This is what appears to me, but I
can not now enlarge any farther on this system. I
have said enough to shew that the subject is very ex-
pensive, and deserves an attentive examination.

XXXIX. *Of*

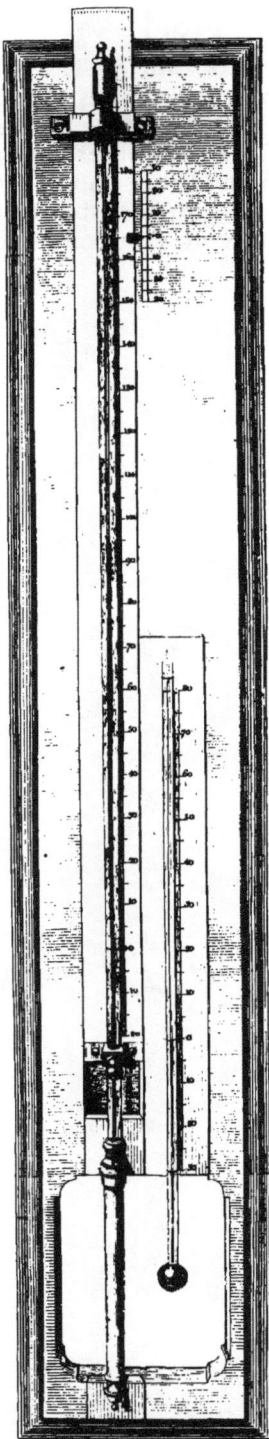

Fig: II.

Fig: I.

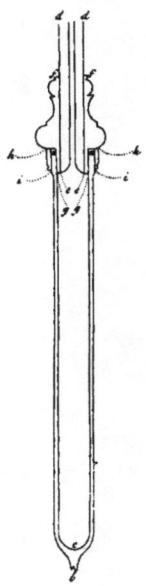

Hygrometer.

www.ingramcontent.com/pod-product-compliance
Lightning Source LLC
Chambersburg PA
CBHW022012190326
41519CB00010B/1494